Grampie

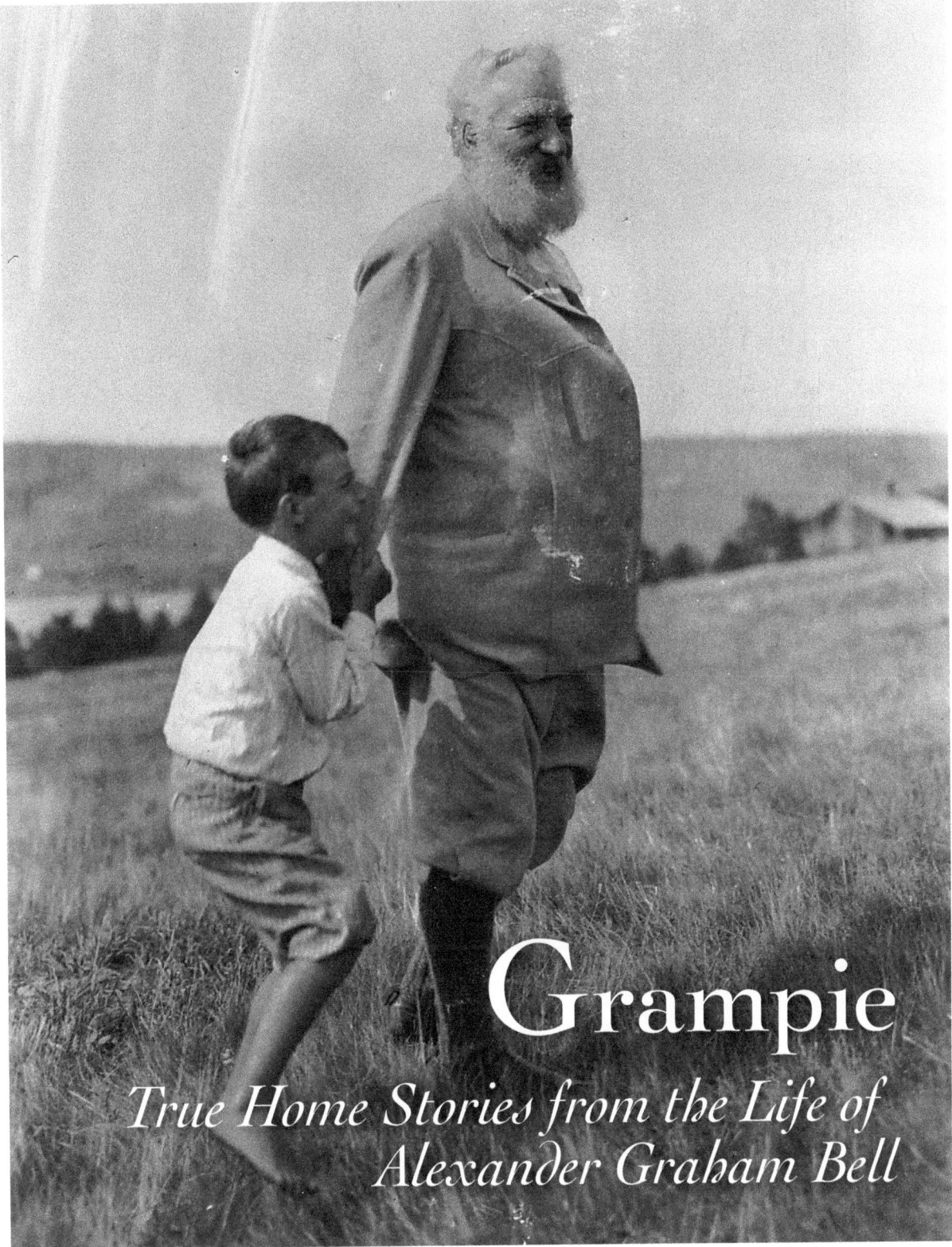

Grampie

True Home Stories from the Life of
Alexander Graham Bell

by Carol Lauritzen *and* Laurel Mathewson

Grampie

True Home Stories from the Life of Alexander Graham Bell

by Carol Lauritzen *and* Laurel Mathewson

wake-
robin
PRESS

Wake-Robin Press
2019

Alexander Graham Bell

Today the telephone lines were silent in memory of my grandfather. For one full minute across America the telephone operators stopped working, and no calls were made or received. The silence of the lines matches the quiet sadness of my heart.

Today my grandfather was buried on a blustery hilltop in Nova Scotia. It was a simple ceremony for such a great man, but we know it is just as he wanted.

After the burial service, we gather inside the family house. I take out the photograph albums and look through the pictures from long ago, when I was "Melly" and he was "Grampie."

From the time I was a baby, I spent summer months with my grandparents in Nova Scotia, in a place called "Beinn Bhreagh," which means "Beautiful Mountain" in Gaelic. Grampie liked to escape the heat of Washington, D.C. in the summertime—and so did I! Surrounded by trees, meadows, and the waters of Bras d'Or Lake, we explored on foot and on horseback, by sailboat and by houseboat.

Beinn Bhreagh

Bell with baby Melville on a horse

Grampie was a scientist through and through—even when at play. On the beach with us children, he didn't just swim and build sandcastles. He demonstrated the position of our solar system in space. He'd place a huge stone on the sand to represent the sun, then pace off the distances from the sun to different planets, putting down smaller stones of varying sizes to represent them—a small stone for Mercury, a slightly larger one for Earth, and a much bigger one for Jupiter. "Where should we put the stars?" I asked. "Oh, they're not on this beach," he answered. "Look over there! The nearest star would be so far from the planets that it would be out of our sight, all the way across Bras D'Or Lake!"

I come across a picture of Grampie at his desk and I smile, remembering the candy jar that sat above it, just for me and my cousins. We learned early that if we went into the study and *asked* for a piece of candy (even politely!), he wouldn't give it. "Oh, no," he'd say, "not now." But if we pretended we were there to see how he was doing and stood there quietly for a moment, he would *always* give us the candy. He called his grandchildren "the candy fiends," especially on afternoons when we interrupted his work three or four times "pretending" not to want candy!

Bell walking with granddaughter Gertrude

Bell and Melville crossing a bridge

Grampie himself had quite a sweet tooth, but when he was older he wasn't supposed to eat much sugar. During the day, Grammy made sure that he didn't eat any unhealthy food, just as the doctor ordered. "Remember the doctor's orders, dear," she'd say. But sometimes, during the night, someone in the family would wake to the sound of noises in the house and get up and notice a light. It was Grampie, sneaking into the kitchen to eat whatever his heart desired. More than once he was caught, cake in hand!

I helped him sneak a few treats, too, though not on purpose. While walking me home from school, he would ask, "Melly, would you like to stop by that bakery and get a snack?" Then he'd buy me a pastry—and one for himself! When we got home, I avoided Grammy's eyes just in case she'd know what had happened from the look on my face. I didn't want to be the one to give away his secret!

The days when my grandfather walked me home from school were lucky ones—and not just because of the pastries. He usually had us walk home from school alone because even when we were small, Grampie encouraged independence. Once, when it was time for me to visit, he decided that I should make the final leg of the trip to the summer house in Nova Scotia, Canada, by myself. My father agreed to the plan and said good-bye to me at his hotel, smiling and waving as I began my walk down the sidewalk to catch the streetcar. I didn't have any specific directions and I was afraid, remembering what happened when Graham, my brother, was encouraged to do the same thing on an errand in our hometown of Washington, D.C. He didn't come home for hours! He missed his streetcar and didn't have enough money for another ticket, so he walked for miles to get back to our house. Grampie just laughed and said it was good for him. I feared the same fate: I'd wander Canada, lost, for months! Thankfully, I sent a telegram (my first) to tell him when I would arrive and got on the right train. Grampie was waiting for me at the station.

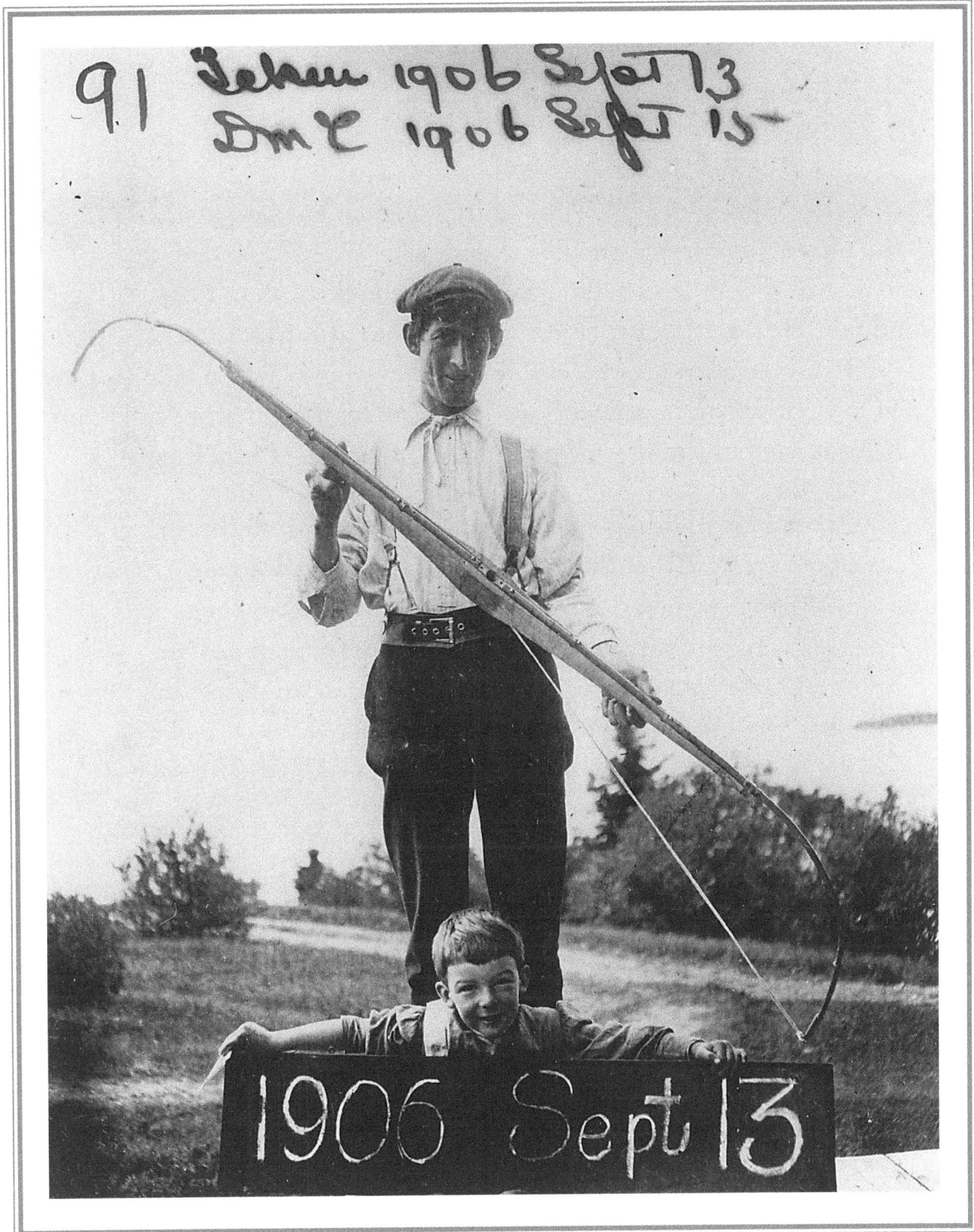

Melville holding a sign below a propeller

My mother comes into the room and hands me hot tea, glancing at the photo album. She sits down and asks, "What are you remembering?" When I tell her, she smiles and gazes out the window. "He was that way with me and your Aunt Daisy, too: Playful and curious; unpredictable and determined." She tells stories from her childhood, stories I have heard many times before but want to hear again and again, especially today. She called him Papa.

"One of my earliest memories," she says, "is the time Papa told me that Santa Claus wanted to talk to me on the telephone. I was beginning to wonder if there really *was* a Santa Claus, so I was very surprised. Our telephone was the old-fashioned box kind on the wall, and I had to stand on a chair to reach it. A jovial voice came over the telephone and said, "Is this Elsie?" When I answered "yes," the voice said, "Well, this is Santa Claus."

"Is it *really* Santa Claus?" I asked.

"Yes, it is really Santa Claus."

"Then where are you?" I demanded.

"I'm up in Greenland," he said.

"Have you *really* got reindeer, and are you coming down here?" I said.

"Yes, I really have reindeer and I'm getting my packs ready to come down and bring presents to good little girls. What do you want me to bring you?"

By that time, he had convinced me that he really was Santa, so I told him in great detail all of the things I wanted. When I got down from the chair in a daze, I turned to Papa, who had been standing beside me. "Isn't it wonderful to be able to talk to Santa Claus?" I said. Still amazed, I said to him, "My, the telephone is a wonderful thing. I wonder who invented it. Do you know, Father?" I looked up to see an amused smile on his face, and then he answered: "I did."

Bell and Gertrude

"There was never a dull moment at the dinner table with Papa," my mother told me. "Someone either tried an experiment, shared tidbits of knowledge from school, or they would have to get the encyclopedia to find a piece of interesting information to share with everyone at the table. One evening, I told Papa that the world was round. He took an orange in his hand. 'Round!' he said. 'Do you mean it is round like this orange? What happens to the people on the underside? Do you mean to say that all those people are standing on their heads and hanging by their toes?' 'I don't think so,' I said. 'But how could people possibly stick to the underside of the world?' he continued. 'Why wouldn't they fall off? …' 'I don't know,' I said. And he kept asking questions that I simply couldn't answer. Eventually I grew so exasperated that I blurted out, 'Oh—God wills it that way!'"

My mother tells me one story I had nearly forgotten: One time, when Papa was in the midst of his kite experiments, two distinguished visitors came to spend a few days with him. One was Professor Langley, secretary of the Smithsonian Institution. The other was Professor Simon Newcomb, astronomer and mathematician. The two men didn't get along well at all, so Papa hoped they'd visit separately. No such luck: Their visits overlapped. One night they were all sitting at the dinner table together and the men began arguing about the laws of gravity. In the midst of the heated debate, Professor Langley mentioned the fact that a cat always lands on its feet. "Not so!" Professor Newcomb retorted. "It all depends on the way the cat is held," he said. "The cat cannot turn over in the air when it's dropped from an upside down position. It is mathematically impossible!" Then Papa piped up. "Well, let's try it!" he said. They got up from the table, found a kitten, and went outside to the veranda, where there was a ten-foot drop to the ground. They scurried to get a mattress and pillows to break the kitten's fall, and then all stood in a circle to watch. The kitten was dropped from upside down—and it landed on its feet! Newcomb insisted that they drop the kitten again and again and again—and each time it landed on its feet. Of course, this only led to a new debate when they sat down again around the table—why did it land on its feet?

Melville and Bell with a spinner

Bell with three granddaughters on a dock

I laugh at the idea of dropping a kitten from a balcony in the middle of a dinner party, but it doesn't surprise me. Grampie's curiosity and passion for homemade, hands-on experiments never dimmed. As the sun sets, my mother recalls stories Grampie told her from his own childhood.

When Papa was quite young, his father presented him with a dead suckling pig. Papa and his brother were part of a club, "The Society for the Promotion of the Fine Arts Among Boys," and they decided to hold a special meeting in their attic so they could show off the pig to the other boys. They arranged boards in rows as seats and placed the dead pig on a table in the front. Papa posed as the distinguished professor of anatomy and began his lecture by stabbing a knife into the belly of the pig. The other boys were very impressed, and it was a grand moment—until a loud, rumbling sound emerged from the pig. The captured air in the pig's body came out, sounding like an angry groan. "It's alive!" the boys yelled in horror. They tumbled over each other as they ran out of the attic, eagerly fleeing toward home. Even Papa, the great fake professor, was too scared to return to the lecture hall to get the pig. His father had to dispose of the corpse for him.

When he was 12, Papa and his brother Melville attempted to make a speaking machine. Melville created the larynx and vocal chords. The larynx was a sheet of tin and he attached a tube as the windpipe to vocal chords made of strips of rubber. It worked by pumping air through a parlor organ into the tube. Papa was in charge of making the mouth and tongue. He started by making a cast from a human skull and used the mold to help replicate the mouth parts, like the lips and tongue. When they put their two creations together, the machine could utter the syllables "mamma" quite distinctly. At the time, they lived in a flat in Scotland with a common stair for many apartments. The boys took their machine out to the stairway. Melville blew as hard as he could and Papa manipulated the lips. The machine yelled out in agonizing cries, "Mamma, mamma, mamma!" From upstairs, a woman opened her door and called out, "My goodness, what is the matter with that baby?" And then they heard the woman begin to walk down the stairs. Papa and Melville sneaked back into their flat, thrilled with their success, and avoided the poor neighbor woman who came to find the screaming "baby."

Bell with granddaughters Gertrude, Lilian and Mabel, 1908

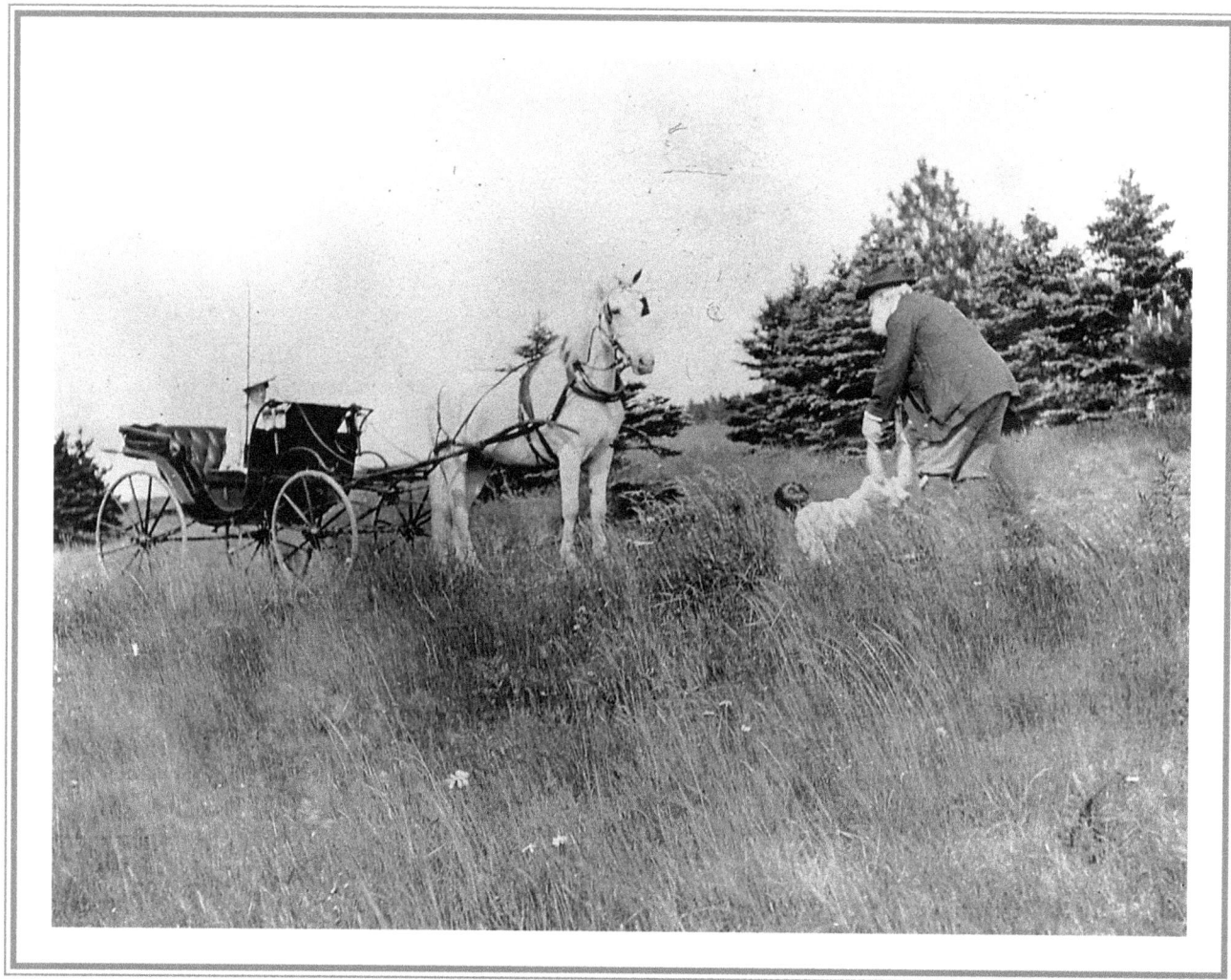

Bell playing with Melville walking on his hands

I smile, thinking of my Grandfather's mischievous, inventive spirit. I wish I could say my curiosity in science got me as far as it did him, but my mischief never resulted in any story so grand. One time, out of curiosity, I poked around in Grampie's den after most of the house had gone to bed. On a table were two small sets of equipment, labeled "Experiment 1" and "Experiment 2." I lifted the dome on the top of Experiment 1 to find a tray of water resting over a strainer underneath. Wanting to see below the tray, I lifted it very carefully, peering over to see creamy, clumpy goop in the strainer underneath. I tilted my head to get a closer look and — oh no! The tray in my hand tilted, too, spilling the water over the creamy goop, and washing most of it away into the bin below. I vowed I'd be much more careful in investigating "Experiment 2," which had a narrow lamp chimney above a regular glass jar. I tried to lift the lamp chimney to see what was inside the glass jar below; it wouldn't budge. *Just a little harder*, I thought. I tugged at it again, and the chimney came loose… but so did the cotton ball that had separated it from the glass jar. Cream poured out of the chimney, spilling into the glass jar and onto the table. I wiped the table with my shirt, put the equipment back together, and sneaked upstairs to bed. I wondered if I'd be caught.

Only later did I find out what happened the next morning. Grampie went to examine the results of his experiments and wrote this in his Home Notes: "Some enterprising individual—name unknown—has already examined them with disastrous results," he wrote in his lab book that day. At the end of the entry, he wrote, "The lesson learned from these experiments is: Don't leave your experimental apparatus where children can get a hold of it. ... Some of them are of an investigating turn of mind. I don't blame them for this, in fact I am rather proud of it, but it may be well to see, that in the future, my experiments are placed out of temptations way." When I read this entry, I remembered the "unknown" culprit: me! So my mischief had been noticed—and he'd even been rather proud of it.

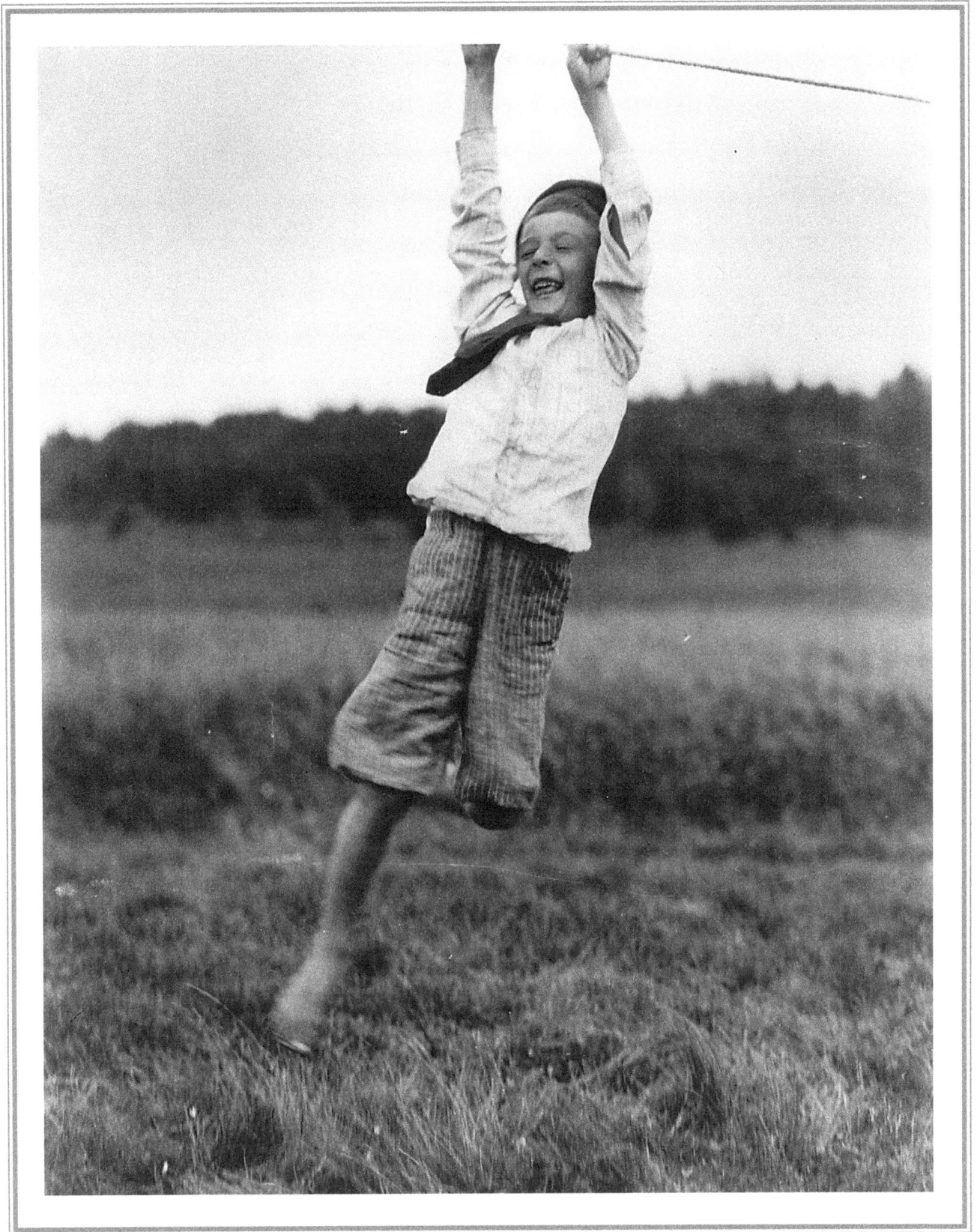

Melville lifted by a kite

Melville and Gertrude with tetrahedron hats, 1908

The summer I was twelve, I spent extra time with Grampie. That year my parents were worried about my behavior ("restless") and my grades in school ("poor"). Maybe because he was worried, too, Grampie designed lots of experiments just for me, using everyday items. What happens when you hang a string in a solution of sugar and water? Do egg shells float in water? In vinegar? Why does a teapot need a spout? Once I wrote my name in wax on an eggshell and dropped it in vinegar. Grampie wanted to know whether I thought the parts with my name would survive *before* I tried it. We didn't just work in the lab, though. Together we sailed around the Bras d'Or lake and made plans to sail all the way to Labrador and New Brunswick someday.

At the end of the summer, Grampie asked my mother and father if I could stay with him and Grammy until Christmas, instead of going back to school. They agreed, so that fall my "school" was life with Grampie. He taught me arithmetic as I helped him finish his scientific study on how long our ancestors lived, and trigonometry while designing a contraption for the fireplace. When beginning the study he told me, "Melly, a table like this, for scientific discovery, must be absolutely accurate. It is necessary to do the whole work twice so as to verify everything." I was extra careful, thinking about how my work was important for science, not just a *grade*. When we compared our work at the end, spotting errors and correcting them, I was proud to see I didn't have too many more mistakes than he did. I even forgot it was the sort of work I might have done at school, too. My "tests" were real projects and experiments—and although it was much more interesting, sometimes I thought a regular test on paper would have been easier!

Bell with Melville writing his name

Bell with a grandchild at a display case

I read to Grampie every day, either from a newspaper or a book. He claimed he needed me to read to him because the print was too small for his old eyes. "Please read this article for me, Melly," he'd say. "It's of great importance to me, and my old eyes just can't quite make it through." I thought I was quite the hero, saving his eyes by reading boring old things that I had no earthly interest in. Later I learned it was a trick. He made me believe that his eyesight was worse than it was so I'd practice reading! Some nights, though, he pretended to have an interest in the Grimm fairy tales. I think he knew they might be more interesting to me than his news and science magazines.

During our reading sessions, I became interested in becoming
an editor myself. I became perfectly possessed over the notion of
having a magazine all my *own* so I started publishing the "Wild Acres
Weekly." I found out that an editor doesn't have to write everything
in the magazine, just find other people to write or use things that are
already written. Grampie suggested that I put some poems in each
issue of my magazine. I *hated* poetry, but he assured me that others
would like to read it. Grandfather suggested one poem in particular
about the number of days in a month, because he thought it would be
beneficial to some of my young cousins. My typing wasn't very good
so I had to type and retype that page until it was orderly enough
to be published. And what do you know, when Grandfather asked
me to recite that poem a few days later, I had learned it by heart—
another trick!

When I see a picture of me that summer with my hair cut completely off, I remember how much I wanted to please Grampie and how determined I was to become the boy he knew I could be. Back then, I had a terrible habit of fiddling with my hair, twirling it 'round and 'round and twisting pieces with my fingertips. Grampie made clear that there was no harm in fiddling with my hair, in itself, but since I did it *hundreds* of times a day, I ought to check the habit. I tried to stop on my own but I had no luck. Next, I asked him to *tell me* whenever he saw me touching my hair so I would let go. He did that for a while, but eventually stopped because he thought it made me too self-conscious. What was I to do? Ever the optimist, he suggested I find some project to keep my hands busy, instead. I couldn't think of anything! Finally, I got so fed up that I went into town by myself and had the barber cut off all my hair. When I returned to the house, I announced to Grampie, "I can't fiddle with it anymore it if it's not there!"

When I tell my mother this story, her eyes grow wide and she tells me to wait. She leaves the room for a few moments, and returns with a paper in her hand. "I was reading old letters from Papa yesterday and came across this. Your haircut made quite an impression on him, too."

Grampie wrote: "Dear Gil and Elsie,

"I have tried to encourage Melly to be self-reliant and overcome his spirit of timidity. Whatever the cause, I can see a marked improvement. You remember, for example, how, only a short time ago, nothing could induce him to ride on his pony for fear of falling off, or being kicked. He has quite overcome this fear and has been riding his pony every day. He even likes to gallop and shows his bravery by leaving go of the reins and holding his hands up. Such an exploit caused him grief the other day. His foot lost the stirrup, he rolled off onto the ground, and the pony stepped on his foot. Only a short time ago such an accident would have led him to the resolution that he would never, never, never ride the pony any more. He was pretty well shaken up, but not much hurt. I came to the conclusion that his riding days were over, for this season at least; but no, he went right on as though nothing had happened; and yesterday rode all the way into Baddeck, and back again, on his pony, to get his hair cut. There is no question about it, he is getting braver. This is surely enough for the present, and I can only say that I love the boy and feel proud of him, and must thank you and Elsie for letting me have him here."

Tears drop to the note as I read it again. *He* thanked my parents for letting me stay? I always felt it was the other way around—that we ought to have thanked *him* for taking me under his wing.

Bell and Melville, hand in hand

Bell and Melville on the boardwalk

As I finish leafing through the photo album, I look around at the big house and out the windows to the familiar sights of Beinn Bhreagh. It is here, with Grampie, that I learned to seek knowledge with passion and joy, with both determination and a twinkling eye, ready to work and to laugh.

Today the world mourns the loss of a great scientist and inventor, because my grandfather is Alexander Graham Bell, famous for inventing the telephone. That's how most people think of him and how most people will remember him. But not me. For me, he will always be Grampie.

Authors' Note:

*Melly is Melville Bell Grosvenor who, as an adult,
became the editor of* National Geographic *magazine.
Although we invented the narrative voice, all of the stories in this book
are based on information from original letters, Bell's Home Notes,
and lab notes archived in the Alexander Graham Bell
Family Papers collection at the Library of Congress
in Washington, D.C. and in the Alexander Graham Bell
National Historic Site of Canada in Baddeck, Nova Scotia.*

Examples of Bell's Simple Experiments

Corked Bottle of Hot Water
Apparently loses some of its
Water on Cooling

(From Home Notes 1915 August 5 p 238)

Take a bottle with a large body a long and narrow neck. Fill it completely full with hot water and then cork it. Place it aside to cool; and in a short time some of the water will, apparently, have disappeared. See Fig 1.

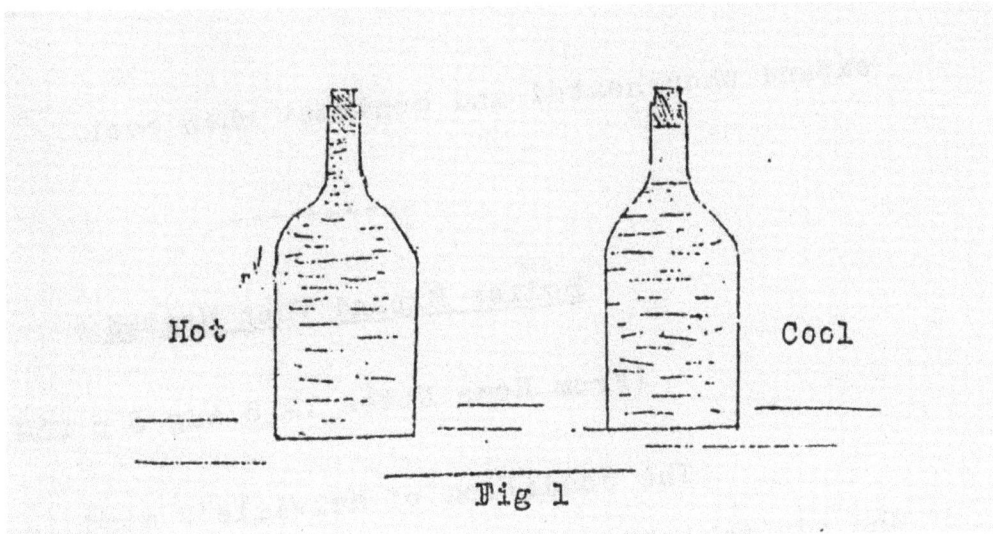

Fig 1

The bottle will be no longer full, although it has been corked to prevent any escape. Water occupies less space when cool than when hot.

To show that no water has escaped have the bottle weighed; and it will be found that it weighs just as much when cool as when hot. This suggests the advisability of weighing things before directing attention to expansion and contraction caused by heat.

Experiment 2

Fill a shallow dish with water (a soup plate will do).
Than take an ordinary fruit jar and drop a lighted
piece of paper into it.

Quickly turn the jar over; and place it, mouth down-wards,
in the dish of water. The flame will go out; and the water
will then leave the dish and go up into the jar.

Why does the water go up into the jar?   AGB

-----------oOo-----------

1915 July 30 ———— Fri ———— at 33 Haboron Swan

escaping water in a third empty tumbler and the water in the three tumblers will ultimately stand at the same level. If the level seems to be too low, pour water into one of the tumblers and the level rises in all three. You can't fill up one of the tumblers without filling them all, and it is really a very pretty and instructive exp. to take a whole row of tumblers connected by syphons. You can put a syphon in the last tumbler of the row to cause the water to flow away, and you can put the first tumbler under a spigot of running water to prevent the water from all escaping.

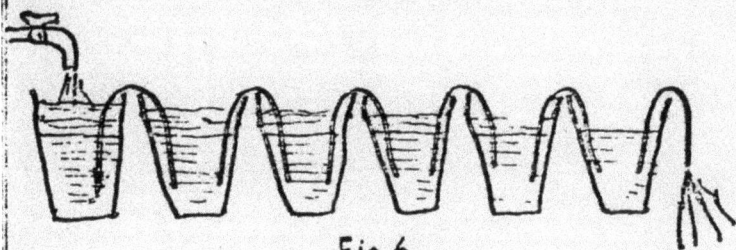

Fig 6.

The water comes in at the first tumbler and flows out at the last but the water can only escape from each tumbler by rising up in the syphon tube above the rim. that is, the water in each tumbler has to flow uphill in order to escape. So that

About the Authors:

Carol Lauritzen *and* Laurel Mathewson *were looking for information about the science experiments Bell developed for household use for potential educational purposes. To their surprise and delight, Bell's "Home Notes" and family letters revealed much more: a rich and tender domestic world, filled with a devotion to growth and learning that went beyond simple experiments. Carol is a retired professor of education who fostered love of learning in countless teachers and students in addition to her two grown sons. Laurel is an Episcopal priest and writer who tries to pass on her own mother's love of scientific observation to her two young children.*

wake-
robin
PRESS

For other titles available from Wake-Robin Press, please visit:
www.wakerobinpress.com

Also available through Ingram, Amazon.com,
Barnesandnoble.com, Powells.com and by special order through
your local bookstore.